BUILDING BLOCKS OF GEOGRAPHY

LAND

Written by Izzi Howell

Illustrated by Steve Evans

WORLD BOOK

a Scott Fetzer company
Chicago

World Book, Inc.
180 North LaSalle Street
Suite 900
Chicago, Illinois 60601
USA

For information about other World Book publications, visit our website at **www.worldbook.com** or call **1-800-WORLDBK (967-5325)**.
For information about sales to schools and libraries, call 1-800-975-3250 (United States), or 1-800-837-5365 (Canada).

Library of Congress Cataloging-in-Publication Data for this volume has been applied for.

Building Blocks of Geography
ISBN: 978-0-7166-4275-6 (set, hc.)

Land
ISBN: 978-0-7166-4282-4 (hc.)

Also available as:
ISBN: 978-0-7166-4292-3 (e-book)

Printed in India by Thomson Press (India) Limited, Uttar Pradesh, India
1st printing June 2022

Acknowledgments:
Writer: Izzi Howell
Illustrator: Steve Evans
Series advisor: Marjorie Frank

Developed with World Book by White-Thomson Publishing LTD

www.wtpub.co.uk

TABLE OF CONTENTS

There is a glossary on page 40. Terms defined in the glossary are in type **that looks like this** on their first appearance.

EARTH'S CRUST
Hi there!
I'm Land! I'm also known as
Earth's crust.
The crust is the outermost shell of the planet. I cover the whole Earth!
Yes, even the parts that are underwater!
Earth's crust is made of rocks and minerals.
Some are worn to tiny pieces of soil or sand ...
... while others, like diamonds and granite, are hard and not easily broken.

Think of Earth's crust like the crust that covers the outside of a loaf of bread!
Like a loaf of bread, there's more to Earth hidden underneath the surface! The crust is just one of four layers that make up our planet.
THE INNER CORE
THE OUTER CORE
THE MANTLE
THE CRUST
The hot rock that makes up much of the mantle can move slowly. The crust floats on top of the mantle—just as a boat floats on the ocean.
The outer core is mostly made of the metals iron and nickel. It's so hot there that the metals have melted into liquids!
Scientists believe that the solid inner core may be hotter than the surface of the sun, with temperatures reaching 16,000 °F (9000 °C)!

Some parts of Earth's crust are covered with water.
This land is known as the sea floor, or **oceanic crust.** It looks just like the land on the surface, with mountains, valleys, and flat plains.

Other parts are above sea level. It is called **continental crust.** This is the land that we live on.

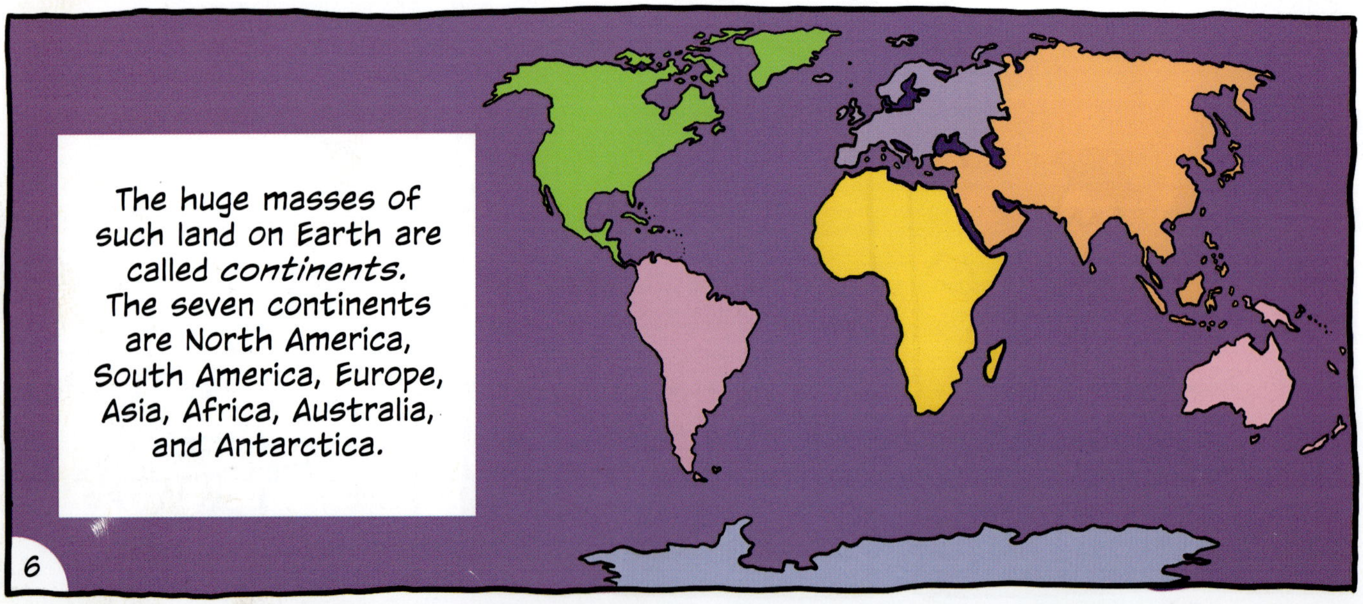
The huge masses of such land on Earth are called *continents.* The seven continents are North America, South America, Europe, Asia, Africa, Australia, and Antarctica.

We wouldn't be able to survive without the land on Earth!
We grow plants for food in its fertile soil.
Many animals depend on plants for food, too...
...and other animals depend on those animals for food!
We gather such resources as rock, metal, and precious stones by digging into Earth's crust.
Such fuels as coal, oil, and natural gas also come from the deep layers of Earth's crust.
Earth's crust provides much of what humans, other animals, and plants need to live! What on earth would we do without land?!

LANDFORMS

Topography is a description of the kinds and arrangements of natural features of an area. These are some of the amazing land formations that are part of Earth's topography.

PEAK
VOLCANO
MOUNTAIN RANGE
MOUNTAIN
WATERFALL
VALLEY
PLATEAU
BUTTE
LAKE
RIVER
OASIS
MESA
MOUTH
BAY
DUNES
PENINSULA
ISTHMUS
CHANNEL
GULF
CAPE
DELTA

The land on Earth is constantly moving and changing, although it usually happens far too slowly for us to see.
These movements and changes create the different landforms you've just seen.
Earth's crust is divided into roughly 30 huge pieces called **tectonic plates.** These slowly move towards, away from, and past each other.
Bye!
See you!
Mountains form at the boundaries between tectonic plates. Some mountains were created when two tectonic plates crashed into each other, pushing up the land in the middle into tall peaks.
Ta-dah!
Other mountains formed when tectonic plates moved away from each other. This motion pushed up a large block of crust, leaving flat **basins** on either side.
FLAT BASIN

Mountains have steep slopes and are much higher than the land around them.
Cheer up, we can't all be number one!
Some mountains stand alone, while others are part of long mountain ranges.
Now that's an angry mountain!
Some mountains are volcanoes in disguise! Many volcanoes form near regions where one tectonic plate is pushed under another.
Not all volcanoes are active. Dormant volcanoes are unlikely to erupt again ...
... while extinct volcanoes will never erupt again. Phew!

Oh, hey **Erosion!** You change the land too, right? You create amazing landforms!
Aww, thanks! But it's not all me.
I wear away rock with the help of some special friends!
Wind!
Water!
Gravity!
Ice!
The movement of Earth's crust, along with the forces of erosion and **weathering,** can cut down through the ground to create deep valleys and canyons ...
Woooo!
... level flat plains ...
... and form high, flat-topped **plateaus** and mesas with steep sides. A plateau is much larger than a mesa.

There are also some pretty cool landforms along the coast and in water.
No need for a boat here! A narrow land bridge called an isthmus connects these two areas.
ISTHMUS
A peninsula is an area of land that is nearly surrounded by water—and is connected by an isthmus to a larger body of land. Some are tiny strips of rock, while others are massive, making up entire countries.
An island is a piece of land that is totally surrounded by water. Some islands are millions of years old, while others are brand-new!
Land! Over here! Didn't you see me waving?
Oh, hey Ocean! Great to "sea" you!
New islands can be formed by underwater volcanic action, by the uplift of land as tectonic plates collide, or by piles of sand deposited through the process of erosion.
A group of islands that form in the same area is known as an archipelago. The country of Japan is an archipelago made up of four large islands, plus over 3,000 smaller islands!
JAPAN

NORTH AMERICA

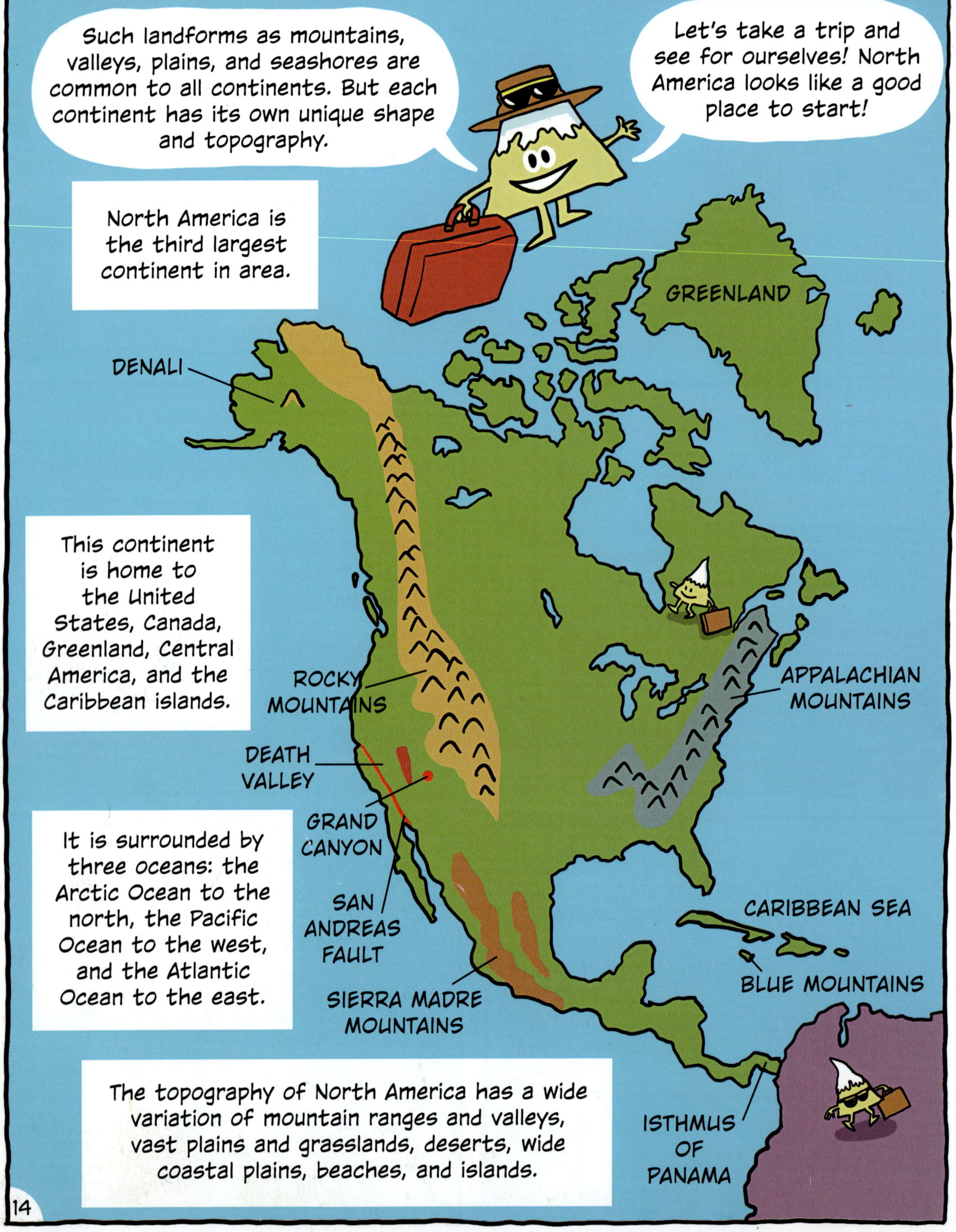

Say cheese!
The Rocky Mountains are the largest mountain system in North America. They stretch for 3,000 miles (4,800 kilometers) down the west of Canada and the United States.
Don't look down! We're at the peak of Denali, the tallest mountain in the USA and North America. This Alaskan mountain, part of the Rocky Mountain system, rises 20,310 feet (6,190 meters) above sea level!
The Sierra Madre mountain system is found around the edges of a huge plateau in the center of Mexico.
There are smaller mountain ranges in Central America and the Caribbean islands. Coffee is grown on the slopes of the Blue Mountains in Jamaica.
Mmmmm!

What other landforms can we see in North America? Ooh, I know!
The canyon is about 1 mile (1.6 kilometers) deep.
The breathtaking Grand Canyon in Arizona, USA, was formed when the Colorado River carved through the layers of rock in the Colorado Plateau over millions of years.
The giant fracture in California, USA, known as the San Andreas Fault, marks the place where two giant tectonic plates meet. The movement of the plates often creates earthquakes.
No, the thermometer isn't broken! The highest temperature ever recorded on Earth, a sweltering 134 °F (57 °C), was taken here, in Death Valley, in the Mojave Desert, USA.
The Great Plains is a large area of flatlands in North America. Much of it is covered in a type of grassland called prairie. This land is very good for growing crops, such as wheat.

North America is home to the largest island in the world-Greenland!
Politically, this icy island is part of the Kingdom of Denmark, but it has its own government.
Question time! Which US state isn't in North America?
The answer is Hawaii! These Pacific islands are considered to be part of North America politically, but geographically, they are part of Oceania.
The islands in Hawaii were created by underwater volcanoes. The layers of lava covering the volcanoes built up over time until the islands sat above sea level.
Many Caribbean islands were also formed by volcanoes. Some of these volcanoes are still active!

SOUTH AMERICA
Time to head down into South America!
South America is connected to North America by a narrow bridge of land called the Isthmus of Panama.
GALAPAGOS ISLANDS
AMAZON RIVER
AMAZON BASIN
ANDES MOUNTAINS
BRAZILIAN HIGHLANDS
ALTIPLANO
South America is the fourth largest continent in area.
OJOS DEL SALADO
ATACAMA DESERT
PARANÁ RIVER
ATLANTIC OCEAN
PACIFIC OCEAN
ACONCAGUA
PAMPAS
It is surrounded by the Pacific Ocean to the west, the Southern Ocean to the south, and the Atlantic Ocean to the east.
The topography of this continent looks somewhat like a bowl–with high mountains or high plateaus around the edges and a large, flat central interior.
PATAGONIA
TIERRA DEL FUEGO
CAPE HORN
SOUTHERN OCEAN

South America has a bit of everything: mountains, huge river valleys, rain forests, grasslands, deserts, and even **glaciers!**
You better bring a drink to the Atacama Desert! This area of Chile and Peru receives less than 0.5 inch (1.3 centimeters) a year, making it one of the driest places on Earth.
Great rivers, such as the Amazon and the Paraná, run through the center of the continent. The Amazon River basin is home to the world's largest tropical rain forest ...
... while the Paraná flows through vast grassy lowland plains with fertile soil. This area is known as the Pampas. The topography here is ideal for grazing livestock and growing wheat and other grains.
Patagonia is the southernmost region at the tip of South America. It is a massive area that includes over 300 glaciers. Glaciers spill down valleys and into the ocean.

The Andes Mountains stretch down the west coast of South America. They are the longest mountain range in the world above sea level, with a total length of around 4,500 miles (7,200 kilometers)!
The world's tallest volcano, Ojos del Salado, is part of the Andes mountain range.
There's no need to panic though–it hasn't erupted in over 1,200 years!
Ojos del Salado isn't the tallest mountain in South America, however. That title belongs to Aconcagua in Argentina which, at 22,841 feet (6,962 meters) tall, is about four-fifths of the height of Mount Everest.
Now we're in the windy Altiplano, a high plateau between two parts of the Andes Mountains. It's one of the highest and largest plateaus in the world.
There are many mines in the Andes because the land is rich in metals, such as copper, gold, silver, and tin. These mountains hold the largest supply of minerals in the world.

The Galapagos Islands are located in the Pacific Ocean, about 600 miles (970 kilometers) off the coast of Ecuador.
The islands and the waters around them form a large marine reserve. The reserve protects a diverse and unique collection of plants and animals, such as this incredible marine iguana!
Hey! I'm a Galapagos tortoise! I only live on the Galapagos Islands. I can grow up to 4 feet (1.2 meters) long!
Wait, penguins? Are we still in South America?
Yes, you are! The Tierra del Fuego islands lie at the very bottom of the continent. The southernmost point of the islands, Cape Horn, is only 600 miles (970 kilometers) from Antarctica!
Many people think that penguins live only in Antarctica. But we've been here since prehistoric times!
We've reached the end of South America, so let's cross the Atlantic to see what we can find!

EUROPE

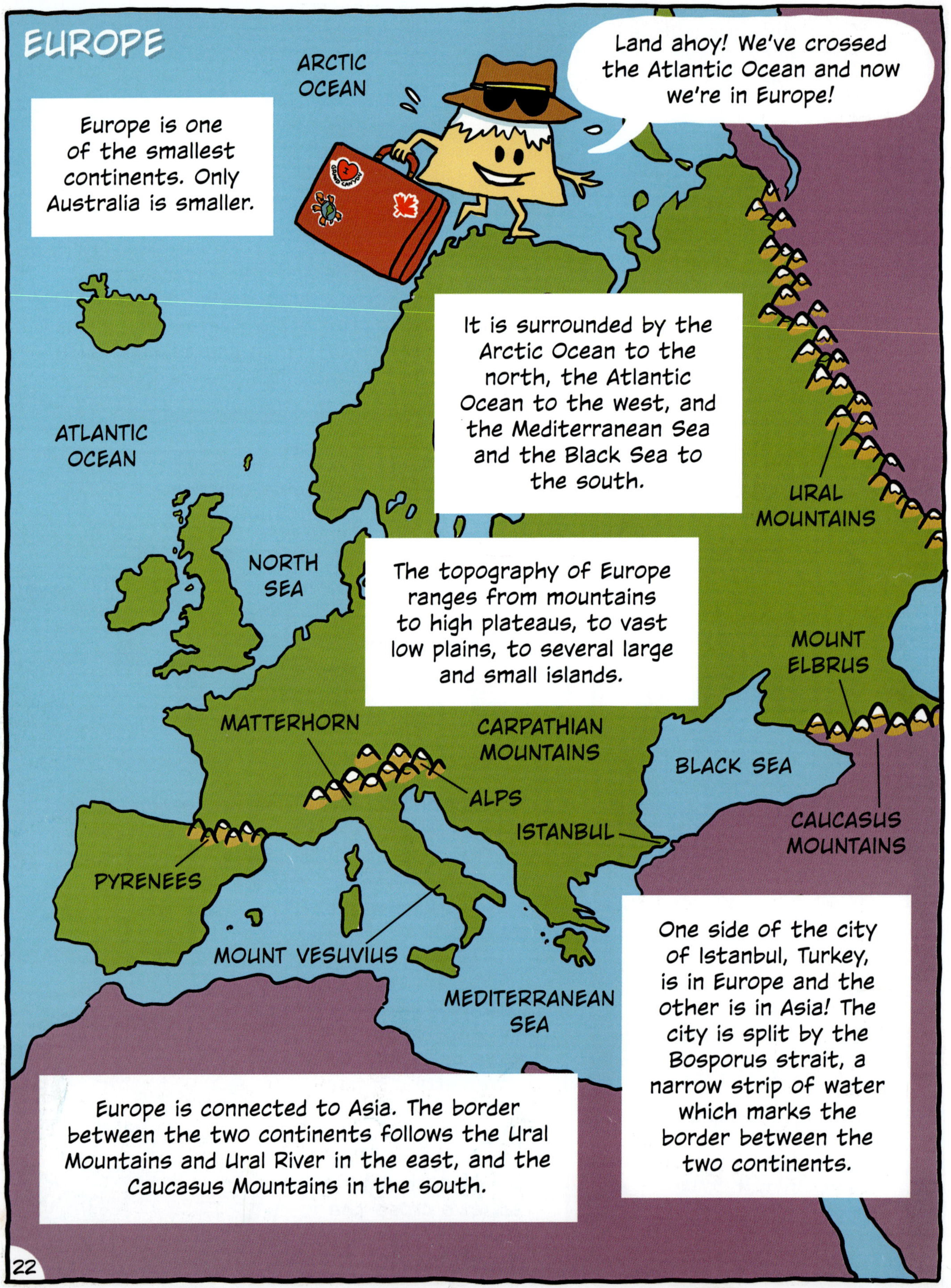

Europe is one of the smallest continents. Only Australia is smaller.

It is surrounded by the Arctic Ocean to the north, the Atlantic Ocean to the west, and the Mediterranean Sea and the Black Sea to the south.

The topography of Europe ranges from mountains to high plateaus, to vast low plains, to several large and small islands.

One side of the city of Istanbul, Turkey, is in Europe and the other is in Asia! The city is split by the Bosporus strait, a narrow strip of water which marks the border between the two continents.

Europe is connected to Asia. The border between the two continents follows the Ural Mountains and Ural River in the east, and the Caucasus Mountains in the south.

Some people say that Europe is a peninsula of peninsulas! The entire continent is a peninsula of Eurasia, which is the name of the area covered by Europe and Asia.
There are also large individual peninsulas within Europe.
The Italian peninsula looks a bit like a boot with a heel!
Europe has an uneven coastline, with many large bays and headlands.
These long, deep inlets of water along the Norwegian coast are known as fiords.
The islands of the United Kingdom and Ireland were once connected to mainland Europe by a land bridge. They were cut off from the continent around 8,000 years ago by rising sea levels.
Oh no!

Yikes! I forgot how many goats are raised in these mountains! I'm here in the Alps, which form the largest mountain system in Europe.
Most of the mountains in Europe, including the Alps, are found across the northern and southern regions of the continent.
In the center of Europe, there are flat plains with fertile farmland. This farmland supports a wide variety of crops.
Europe is also rich in other resources, such as timber from the forested areas of Scandinavia, Germany, France, and Belgium ...
... and oil and natural gas from the sea floor of the North Sea.

Hey, how would you like a travel buddy?
You bet, Erosion! Hope you've brought your climbing boots! There are tons more mountains to explore.
Check out this pyramid! ... are we in Egypt?
No, we're still in the Alps! The Matterhorn naturally has an almost perfect pyramid shape! It was carved into this shape over time by the movement of glaciers.
Tell me we're almost at the top!
Just a little farther! Mount Elbrus in Russia is the tallest mountain in Europe, after all. It measures 18,510 feet (5,642 meters), so getting to the peak is no walk in the park!
I'm out of here!
Oh, that's just Mount Vesuvius in Italy! It's the only active volcano on the European mainland. It famously destroyed ancient Roman cities when it erupted in A.D. 79, and it's still going strong today!

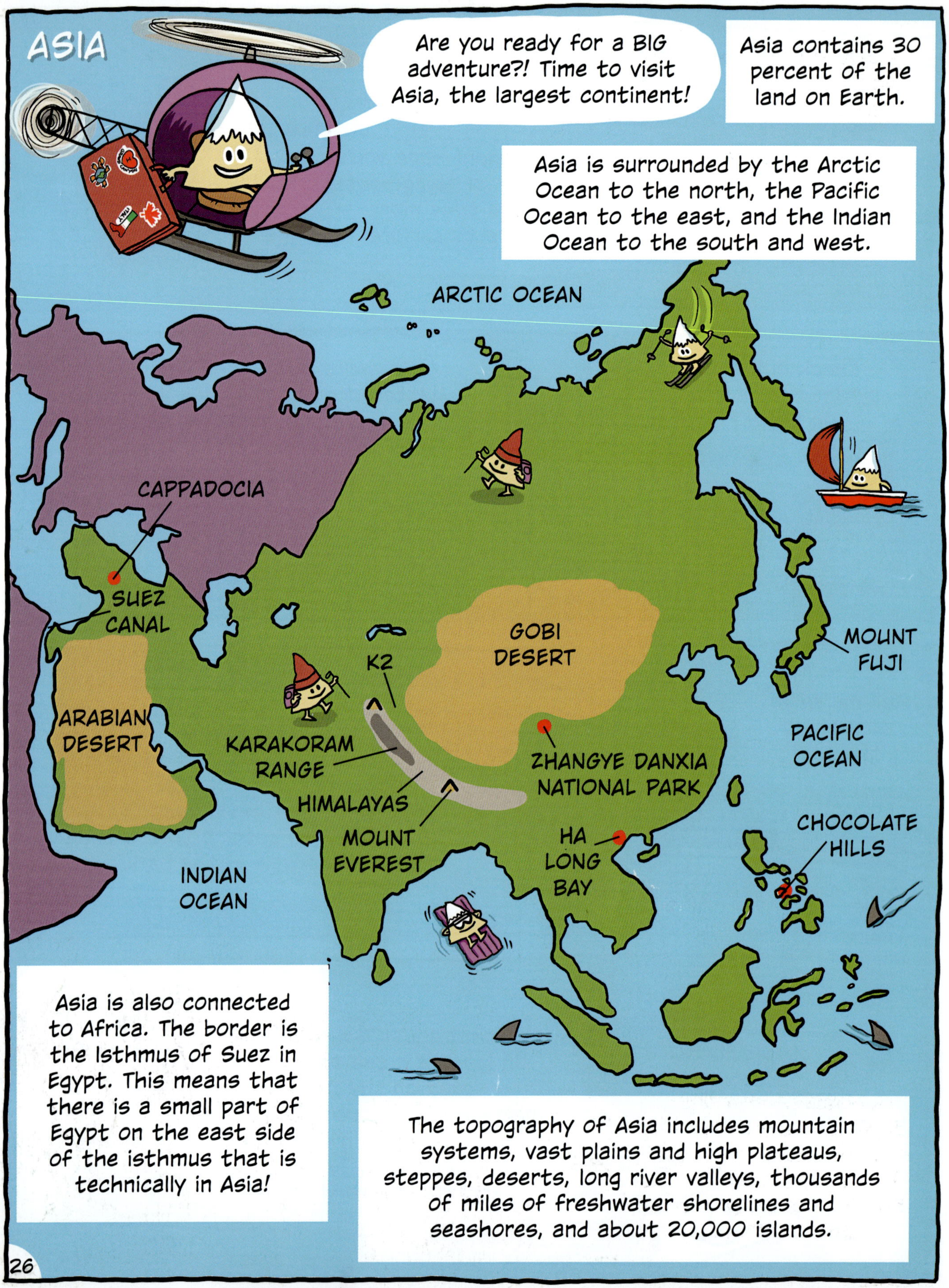
ASIA
Are you ready for a BIG adventure?! Time to visit Asia, the largest continent!
Asia contains 30 percent of the land on Earth.
Asia is surrounded by the Arctic Ocean to the north, the Pacific Ocean to the east, and the Indian Ocean to the south and west.
ARCTIC OCEAN
CAPPADOCIA
SUEZ CANAL
ARABIAN DESERT
K2
GOBI DESERT
KARAKORAM RANGE
HIMALAYAS
MOUNT EVEREST
ZHANGYE DANXIA NATIONAL PARK
HA LONG BAY
MOUNT FUJI
PACIFIC OCEAN
CHOCOLATE HILLS
INDIAN OCEAN
Asia is also connected to Africa. The border is the Isthmus of Suez in Egypt. This means that there is a small part of Egypt on the east side of the isthmus that is technically in Asia!
The topography of Asia includes mountain systems, vast plains and high plateaus, steppes, deserts, long river valleys, thousands of miles of freshwater shorelines and seashores, and about 20,000 islands.

I always feel at home in Asia. Maybe it's because it has more mountains than any other continent on Earth!
The Himalaya were created when the Indian subcontinent crashed into Asia about 50 million years ago. The tectonic plates are still crashing, which makes the mountains grow about 2 inches (5 centimeters) taller every year!
Puff! It's a long way to the top of Mount Everest! Its peak is about 5 1/2 miles (8.85 kilometers) above sea level, making it the highest mountain on Earth. It is located in the Himalaya.
K2 is so high that its peak is usually hidden by clouds! This mountain, which is the second highest on Earth at 28,250 feet (8,611 meters), is found in the Karakoram Mountain Range.
Surprise! Did you know that Mount Fuji is an active volcano? It is located just 62 miles (100 kilometers) from Tokyo, the capital of Japan. Luckily, it hasn't erupted since 1707.

It's not all mountains in Asia though! Central Asia has many flat areas of grassland with no trees, known as steppes. Bactrian camels are native to this area.
There are also many deserts in Asia. The Arabian Desert covers nearly all of the Arabian Peninsula!
Many important prehistoric fossils have been found in the Gobi Desert in China and Mongolia. These include amazing fossilized dinosaur eggs discovered in 1923.
Siberia, the vast northern part of Asia, is mostly cold, treeless tundra. The tundra is considered to be a kind of desert, even though it's frozen! That's because it receives so little rain!
ITALY
I guess I didn't need to bring my umbrella after all!

You can't visit Asia without stopping by some of its incredible landforms.
These incredible rock formations in Cappadocia, Turkey, are known as fairy chimneys! They were created when wind and rain eroded away volcanic rock. The doors and windows were added when people carved out homes inside!
Just admiring our work!
There are thousands of limestone towers, hills, and tiny islands in Ha Long Bay, Vietnam.
Wow! These rainbow mountains are found in Zhangye Danxia National Park in China. Iron and other minerals have stained the rock, creating beautiful vibrant colors.
Chocolate Hills? No, not that kind of chocolate!
The green grass that covers these hills in the Philippines turns brown during the dry season, making them look like huge mounds of chocolate! Yum!
These weathered limestone forms are the result of erosion.

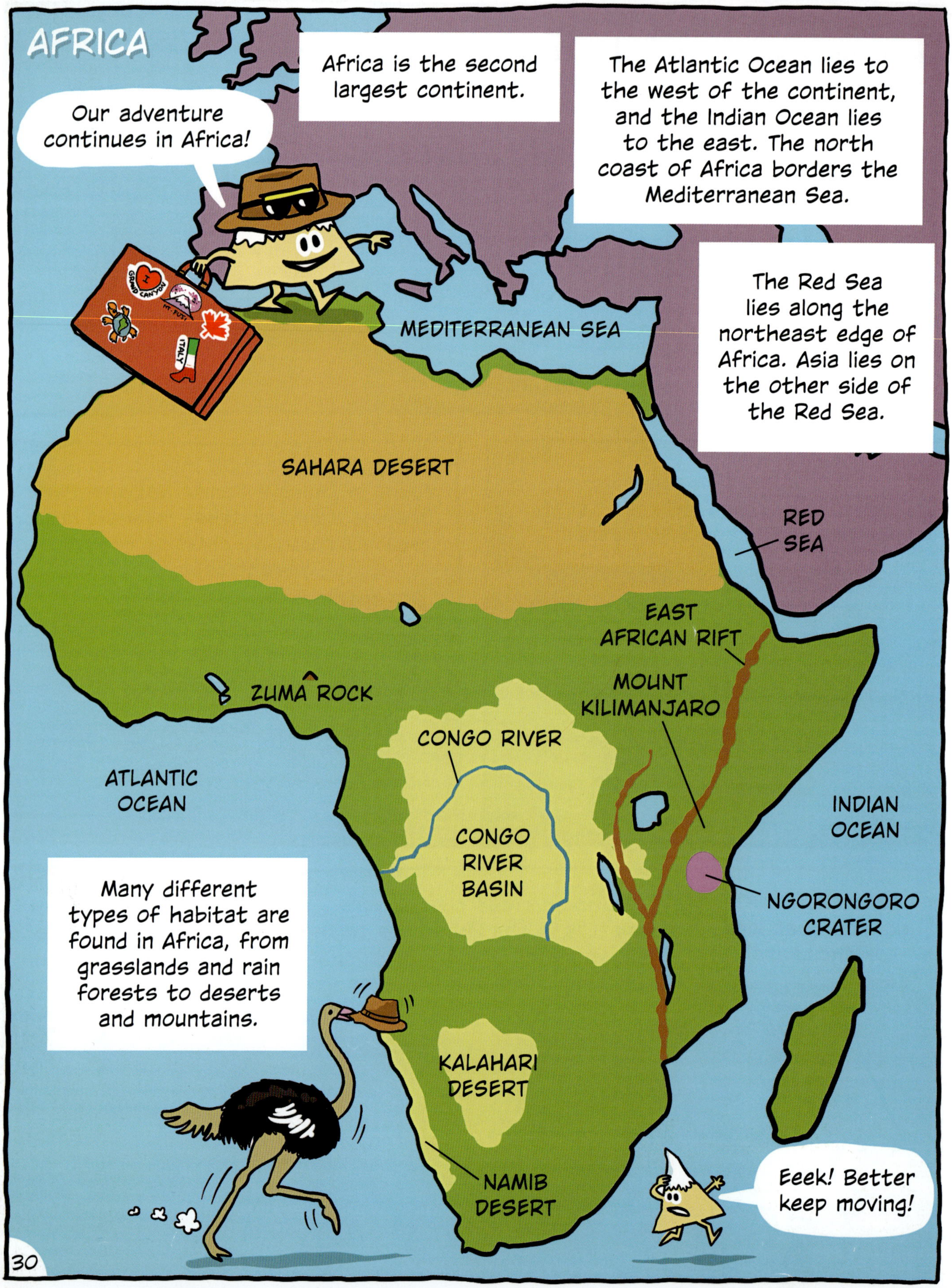
AFRICA
Our adventure continues in Africa!
Africa is the second largest continent.
The Atlantic Ocean lies to the west of the continent, and the Indian Ocean lies to the east. The north coast of Africa borders the Mediterranean Sea.
The Red Sea lies along the northeast edge of Africa. Asia lies on the other side of the Red Sea.
I ♥ GRAND CANYON
MT. FUJI
ITALY
MEDITERRANEAN SEA
SAHARA DESERT
RED SEA
EAST AFRICAN RIFT
MOUNT KILIMANJARO
ZUMA ROCK
CONGO RIVER
ATLANTIC OCEAN
CONGO RIVER BASIN
INDIAN OCEAN
NGORONGORO CRATER
Many different types of habitat are found in Africa, from grasslands and rain forests to deserts and mountains.
KALAHARI DESERT
NAMIB DESERT
Eeek! Better keep moving!

Africa is rich in many different natural resources.
Its lands contain large amounts of **fossil fuels,** such as coal, oil and natural gas ...
...precious metals and gems, such as gold, platinum, and diamonds...
...as well as metals used to make many everyday items. Africa has large stores of copper, zinc, tin, and bauxite, from which aluminum is obtained.

Large areas of Africa are high plateaus ...
... broken up by rocky outcrops, such as Zuma Rock in Nigeria ...
... and the occasional tall mountain! Mount Kilimanjaro in Tanzania is the tallest mountain in Africa.
Even though it lies close to the equator, it's so tall that it has snow at its peak for most of the year. Good thing I brought my hat and scarf!
Watch out! Africa is splitting in two along the East African Rift! This rift valley is a low region formed as the main African tectonic plate pulls away from a smaller tectonic plate to the east.
The Congo River, the world's deepest river, flows from high plateaus in central Africa, across plains and through the lowlands of the Congo Rainforest, to the Atlantic Ocean.
Boo!

Around two-fifths of Africa is covered by deserts. The largest African desert, and the largest hot desert on Earth, is the Sahara Desert. It's about the same size as the USA!
Some sand dunes in the Sahara Desert are 600 feet (180 meters) high. That's taller than the Great Pyramid of Giza in Egypt!
Safari time! The Ngorongoro Crater in Tanzania is a great place to spot some wildlife! This crater is a **caldera**–a huge basin left behind after a massive volcanic eruption.
This is the largest caldera in the world, measuring around 100 square miles (260 square kilometers). That's bigger than the city of Washington, D.C., USA!
Uh-oh, I think it's lunchtime here ... time to move on!

AUSTRALIA

Australia is the only country that is also a continent. It is the smallest continent by area. The Pacific Islands consist of thousands of islands scattered across the Pacific Ocean. The islands that lie close to the mainland of a continent are considered part of that continent. Most Pacific Islands, however, do not belong to any continent. They are part of a vast Pacific region called Oceania.

The largest islands of Oceania are New Guinea and New Zealand's large North Island and South Island. Oceania's small islands number in the thousands.

Sometimes people include Australia in Oceania based on its location. Most geographers do not include it, though, because they consider Australia to be a continent.

Nice to meet you!
The Three Sisters are an unusual rock formation in the Blue Mountains, Australia. Early European settlers gave the Blue Mountains their name because they usually appear in a bluish haze. Fine drops of eucalyptus oil in the atmosphere cause the haze.
Uluru, also known as Ayers Rock, is a giant rock located near the center of Australia. It is sacred to the Anangu people, who are one of the **Indigenous** peoples of Australia. The Anangu people have looked after the land surrounding Uluru for tens of thousands of years. The rock appears to glow red at sunrise and sunset.
I've never surfed a wave like this before!
This unusual granite cliff in western Australia is known as Wave Rock because of its incredible shape. Nice work if I do say so myself!
Millions of years of erosion have carved the rock here into balls. This area of Australia is known as Karlu Karlu or Devils Marbles. It is a conservation reserve to protect a sacred site. Some rocks are balanced on top of each other, while others have split down the center.
Oops!

Australia sits on its own tectonic plate.
The edge of the tectonic plate runs across New Zealand. As a result of plate movement, there are many active volcanoes and earthquakes there. Let's take a look at how this has affected the landscape.

Where are we now? It looks like a lake, but we're actually on a volcano!
A huge caldera was left behind after the huge Taupo volcano erupted in New Zealand about 27,000 years ago. The basin filled with water and ta-dah … Lake Taupo was born!

The Southern Alps and High Country cover most of New Zealand's South Island. The Southern Alps are also called by their Māori name, *Kā Tiritiri o te Moana*. They were formed by the collision of tectonic plates. The highest mountain in New Zealand, Aoraki/Mount Cook, is found here. The Māori name *Aoraki* is usually translated as *cloud piercer*.

Whoa! Whales sometimes swim into the 10-mile-(16-kilometer)-long Milford Sound. This stunning fiord in New Zealand was created by glaciers over millions of years.

We can't visit this region without checking out some of Oceania's many islands. They can be divided into two types-high and low.
High islands, such as New Guinea and the main islands of Fiji and Vanuatu, were formed by volcanic eruptions.
Low islands, such as Kiribati and the Marshall Islands, often form on top of coral atolls. An atoll is a ring-shaped coral reef that builds up around a sinking volcanic island or on the crater of a sunken volcano.
HELLO NEW ATOLL!
VOLCANO
CORAL REEF
It sounds weird for an island to form on a coral reef. But coral skeletons actually make up the limestone rock of the atoll.
As their name suggests, most low islands rise just above sea level. If sea levels go up as a result of **global warming,** these places will start to disappear.
I hope this isn't my last trip to see them.

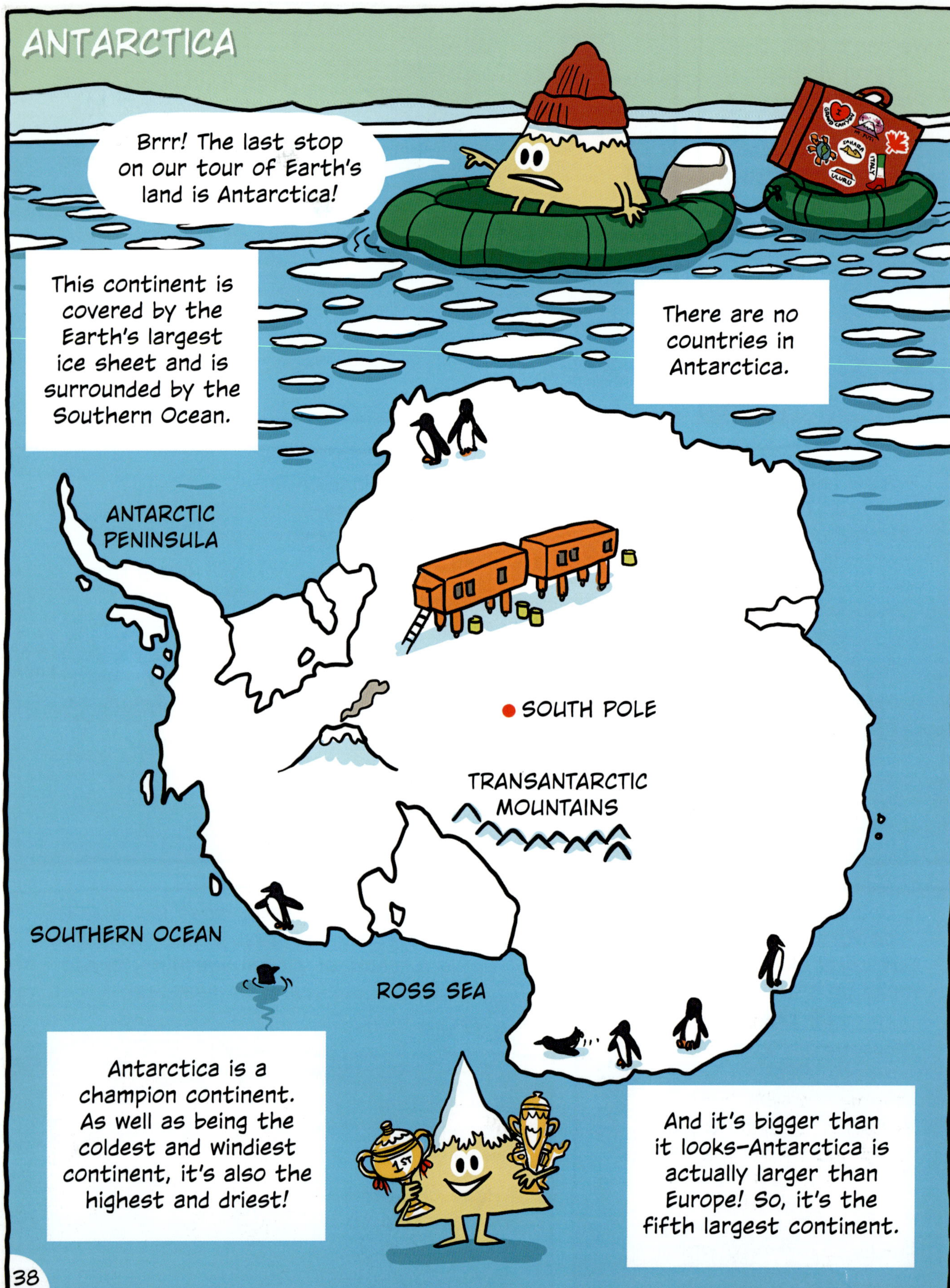
ANTARCTICA
Brrr! The last stop on our tour of Earth's land is Antarctica!
This continent is covered by the Earth's largest ice sheet and is surrounded by the Southern Ocean.
There are no countries in Antarctica.
ANTARCTIC PENINSULA
SOUTH POLE
TRANSANTARCTIC MOUNTAINS
SOUTHERN OCEAN
ROSS SEA
Antarctica is a champion continent. As well as being the coldest and windiest continent, it's also the highest and driest!
And it's bigger than it looks–Antarctica is actually larger than Europe! So, it's the fifth largest continent.

These mountain peaks are glimpses of the land hiding underneath the snow.
If the ice and snow melted away, Antarctica would look similar to any other land, with mountains, valleys, desert plains and other landforms.
The Antarctic Peninsula extends from the north of the continent in an S shape. There are many mountains along the peninsula, which are connected to the Andes Mountains in South America by an underwater ridge.
Wait, where did all the snow go?
Now we're in a dry valley in the Transantarctic Mountains. There is no snow because it is so dry! The little snow that does fall here is blown away by the strong winds.
This post marks the spot of the South Pole! This is the southernmost point on Earth. It is located near the center of Antarctica.
It's a myth that we penguins live at the South Pole. We are sea animals! But we do nest on land. I spend some of my time on the Antarctica Peninsula and shores of the Ross Sea.
Puff! What a trip! I'm exhausted now. What an incredible land we live on!

WORDS TO KNOW

basin an area of land that is lower than its surroundings.

caldera a large depression formed when a volcano erupts and collapses.

continental crust the dry land of Earth's surface that makes up the continents.

erosion the process by which Earth materials are worn away by wind, water, gravity, or ice.

fossil fuels sources of energy that formed from the remains of living things that died millions of years ago. Coal, oil, and natural gas are fossil fuels.

glacier a large river or sheet of ice that moves slowly.

global warming the increase in temperature on Earth.

Indigenous the first people who lived in a particular region.

minerals naturally occurring solid substances in the earth, such as gold, rock salt, graphite, and iron ore.

Oceania a region of the Pacific Ocean that includes numerous small islands, as well as New Guinea and New Zealand.

oceanic crust dark, dense volcanic rock that makes up the land of the ocean floor.

plateau a raised area of relatively flat land.

political having to do with citizens or a government of a country.

tectonic plate a massive, irregular-shaped slab of rock, usually composed of continental and oceanic crust. Earth's surface is made up of about 30 tectonic plates.

weathering the natural process by which rock is broken down into smaller and smaller pieces.